WILLIAM WOODVILLE, M.D. F.R.S.

Author of Medical Botany, &c.

View of the inoculating Hospital at Pancras

Title: **MEDICAL & OFFICINAL PLANTS - VOL. 1 By William Woodville & James Sowerby XIX centuries engravings**
Series realized by Luca S. Cristini. Scientific consulence by Marco Rampinelli.

ISBN code: 978-88-93272285 First edition March 2017
Code.: **MUSEUM-004**, Editorial series code: Darwin's View **DV-003**

Cover & Art Design: Luca S. Cristini & Anna Cristini
MUSEUM is a trademark of Soldiershop publishing, via Padre Davide, 7 - 24050 Zanica (BG) ITALY. www.bookmuseum.it

WILLIAM WOODVILLE - JAMES SOWERBY

MEDICAL & OFFICINAL PLANTS - VOL. I

PIANTE OFFICINALI, MEDICINALI E AROMATICHE

S hown in this series of three books are the complete and original pattern hand-colored engravings plates from the artist James Sowerby's medical plants present in the great work of William Woodville: Medical botany (London : Printed and sold for the author, by James Phillips, 1790-1793). Medical Botany, William Woodville's three volume work of materia medica, was published in monthly installments between 1790 and 1793.

A third edition of five volumes (the same used in our reproduction) was presented in 1832, twenty-seven years after Woodville's death. This publication added descriptions of thirty-nine new plants and was edited and revised by the eminent botanist, William Jackson Hooker (1785-1865).

With this work, Woodville intended to educate medical practitioners about the plants they prescribe and improve upon preceding works by introducing new plants and more detail.

William Woodville was an English physician and botanist who lived and worked in London for most of his life. For his work in botany, he was made a Fellow of the Linnaean Society just a year after the publication of volume one of his Medical Botany.

DARWIN'S VIEW

MUS EUM
BIBLIOTHECA

MEDICAL BOTANY

Containing systematic and general descriptions, with plates, of all the medicinal plants, indigenous and exotic, comprehended in the catalogues of the *materia medica*, as published by the Royal colleges of physicians of London and Edinburgh: accompanied with a circumstantial detail of their medicinal effects, and of the diseases in which they have been most successfully employed.

By William Woodville, M. D. Of the Royal college of physicians, London. Printed and Sold for the Author, by James Phillips, in five volumes (Edition of 1832)

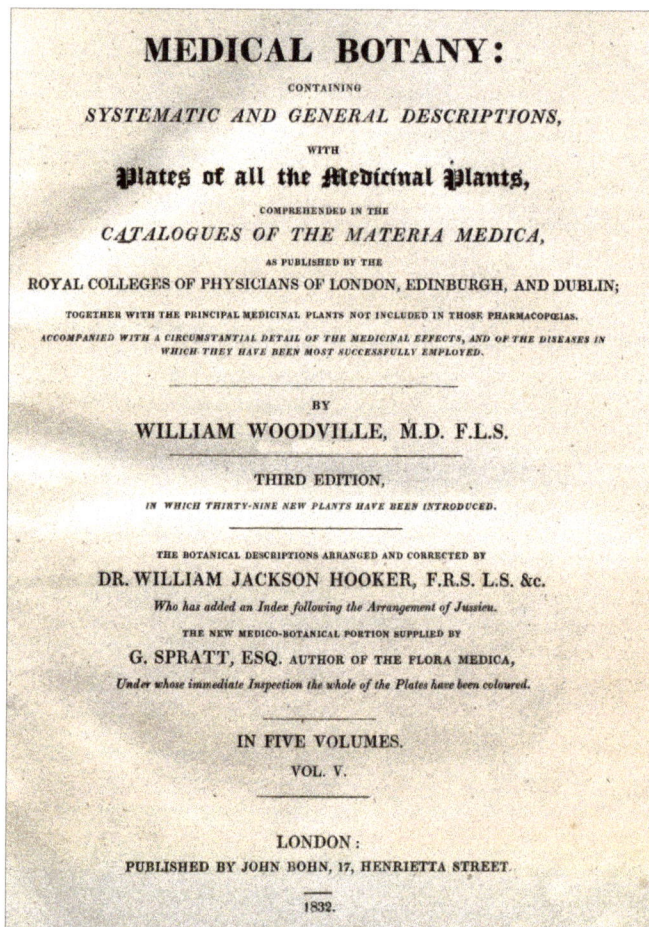

MEDICAL BOTANY:

CONTAINING

SYSTEMATIC AND GENERAL DESCRIPTIONS,

WITH

Plates of all the Medicinal Plants,

COMPREHENDED IN THE

CATALOGUES OF THE MATERIA MEDICA,

AS PUBLISHED BY THE

ROYAL COLLEGES OF PHYSICIANS OF LONDON, EDINBURGH, AND DUBLIN;

TOGETHER WITH THE PRINCIPAL MEDICINAL PLANTS NOT INCLUDED IN THOSE PHARMACOPŒIAS.

ACCOMPANIED WITH A CIRCUMSTANTIAL DETAIL OF THE MEDICINAL EFFECTS, AND OF THE DISEASES IN WHICH THEY HAVE BEEN MOST SUCCESSFULLY EMPLOYED.

BY

WILLIAM WOODVILLE, M.D. F.L.S.

THIRD EDITION,

IN WHICH THIRTY-NINE NEW PLANTS HAVE BEEN INTRODUCED.

THE BOTANICAL DESCRIPTIONS ARRANGED AND CORRECTED BY

DR. WILLIAM JACKSON HOOKER, F.R.S. L.S. &c.

Who has added an Index following the Arrangement of Jussieu.

THE NEW MEDICO-BOTANICAL PORTION SUPPLIED BY

G. SPRATT, ESQ. AUTHOR OF THE FLORA MEDICA,

Under whose immediate Inspection the whole of the Plates have been coloured.

IN FIVE VOLUMES.

VOL. V.

LONDON:
PUBLISHED BY JOHN BOHN, 17, HENRIETTA STREET.

1832.

Medicus omnium Stirpium (si fieri potest) peritiam habeat; sin minus plurium saltem quibus frequenter utimur.   *Galen, Lib. De Antidot.*

PREFACE

In the catalogues of the *Materia Medica*, the productions of the animal and mineral kingdoms bear a small proportion to those of the vegetable.

Though it must be acknowledged that for some time past the medicinal uses of vegetable simples have been less regarded by physicians than they were formerly, which probably may be ascribed to the successive discoveries and improvements in chemistry;

it would however be difficult to shew that this preference is supported by any conclusive reasoning drawn from a comparative superiority of Chemicals over *Galenicals*, or that the more general use of the former has actually led to a more successful practice.

Although what may be called the herbaceous part of the *Materia Medica*, as now received in the British pharmacopoeias, comprises but a very inconsiderable portion of the vegetable world; yet limited as it now is, few medicinal practitioners have a distinct botanical knowledge of the individual plants of which it is composed, though generally, well acquainted with their effects and pharmaceutical uses.

But the practitioner, who is unable to distinguish those plants which he prescribes, is not only subjected to the impositions of the ignorant and fraudulent, but must feel a dissatisfaction which the inquisitive and philosophic mind will be anxious to remove, and to such it is presumed Medical Botany, by collecting and supplying the information necessary on this subject, will be found an acceptable and useful work; the professed design of which is not only to enable at the reader to distinguish with precision all those plants which are directed for medicinal use by the Colleges of London and Edinburgh, but to furnish him at the same time with a circumstantial detail of their respective virtues, and of the diseases in which they have been most successfully employed by different writers.

A distinctive and characteristic knowledge of natural objects mould certainly precede the consideration of their different properties and qualities; and with respect to plants, this knowledge is seldom to be adequately attained by a mere verbal description: accurate delineations therefore, become necessary, and this department is committed to Mr, Sowerby, an artist of established reputation, whole talents are not less conspicuous in the correctness than in the beauty of his designs.

It is justly a matter of surprise, that notwithstanding the universal adoption of the Linnaean system of Botany, and the great advances made in natural science, the works of Blackwell and Sheldrake should still be the only books in this country in which copper-plate figures of the medicinal plants are professedly given; while splendid foreign publications of them, by Regnault, Zorn, and Plenk, have appeared in the space of a very few years. These works however are far from superseding that now offered to the public; for without resorting to the invidious talk of pointing out their errors and imperfections, the author has the satisfaction of having exhibited Icons of several rare and valuable plants, which have never been completely figured in any preceding work whatever: and by subjoining some account of the botanical and medical history of each species, curiosity is more fully gratified, and a double interest is excited in the mind of the student.

Respecting the uses of Simples, the opinion of Oribasius will not be disputed, viz. " *Simplicium medicamentorum,& facultatum qua in eis insunt, cognitio ita necessaria est, ut sine ea nemo rite medicari queat* " and it is a lamentable truth, that our experimental knowledge of many of the herbaceous simples is extremely defective; for as writers on the *Materia Medica* have usually done little more than copy the accounts given by their predecessors, the virtues now ascribed to several plants are wholly referable to the authority of Dioscorides.

It is however hoped that the medical reader will find what relates to this part of the work as complete as the flow progressive state of experience in physic will admit: with this intention, facts and opinions have been industriously collected from various authorities; and those adduced by Professor Murray, and the works of the late Dr. Cullen, have furnished the largest contribution.

The publication of this work in monthly numbers has afforded the author an opportunity of knowing already the sentiments entertained of it, by several Gentlemen of great medical and botanical authority; from whose unsolicited communications he has derived considerable assistance, and for whose friendly suggestions he desires to make his most grateful acknowledgements.

(from the preface of first edition of Woodville work.)

Note on the plates

While today we may look at these pattern plate books as works of art, it is important to remember their initial function was to provide accurate representations of natural phenomena to promote the study of natural history. These notated hand-colored plates were meant to provide guidance in terms of color and shading to others involved in the printing process.

Details regarding the plates: copper plate engravings in black printing ink, hand colored with watercolor, with iron gall ink and graphite inscriptions.

Duplex est dos libelli.

BIOGRAPHIES

William Woodville (1752-1805) was a scientist born into a Quaker family at Cockermouth, Cumberland. He stusies medicine at the university of Edinburgh in Scotland, graduating MD on 12 September 1775. He start his work in European mainland before entering practice in his native country. After he became a famous physician and in 1791 was elected director of the prestigious Smallpox and Inoculation hospitals, St. Pancras in London. With this assignment he appropriated two acres at Battle bridge , near the hospital, where he established a botanical garden, which was maintained at his expenses. In the same year he realize the three volume treatise, *Medical Botany*. In this work , Woodville described, with illustrations and accounts of therapeutic effects all know medicinal plants at this time.

▲ William Woodville M.D. portrait

▼ James Sowerby portrait by Heaphy (1816)

James Sowerby (1757-1822) was an English naturalist and illustrator. Trained at the Royal Academy He gained a reputation as a fine illustrator for his contributions to works on botany and other fields of natural science. His reputation was such that he later was involved in several multi-volumed sets published over several years. His family, including both sons and daughters, were involved in his publication endeavors.

MEDICINAL PLANTS

Medicinal plants have been identified and used from prehistoric times. Plants make many chemical compounds for biological functions, including defence against insects, fungi and herbivorous mammals. Over 12,000 active compounds are known to science. These chemicals work on the human body in exactly the same way as pharmaceutical drugs, so herbal medicines can be beneficial and have harmful side effects just like conventional drugs. However, since a single plant may contain many substances, the effects of taking a plant medicine can be complex.

The earliest historical records of herbs are found from the Sumerian civilisation, where hundreds of medicinal plants including opium are listed on clay tablets. The Ebers Papyrus from ancient Egypt describes over 850 plant medicines. Drug research makes use of ethnobotany to search for pharmacologically active substances in nature, and has in this way discovered hundreds of useful compounds. These include the common drugs aspirin, digoxin, quinine, and opium. The compounds found in plants are of many kinds, but most are in four major biochemical classes, the alkaloids, glycosides, polyphenols, and terpenes.

HISTORY

Plants, including many now used as culinary herbs and spices, have been used as medicines from prehistoric times. Spices have been used partly to counter food spoilage bacteria, especially in hot climates, and especially in meat dishes which spoil more readily. Angiosperms (flowering plants) were the original source of most plant medicines. Human settlements are often surrounded by weeds useful as medicines, such as nettle, dandelion and chickweed. Some animals such as non-human primates, monarch butterflies and sheep ingest medicinal plants to treat illness.

Plant samples from prehistoric burial sites are among the lines of evidence that Paleolithic peoples had knowledge of herbal medicine. For instance, a 60.000-year-old Neanderthal burial site, "Shanidar IV", in northern Iraq has yielded large amounts of pollen from 8 plant species, 7 of which are used now as herbal remedies. The deliberate placement of flowers has been challenged. Paul B. Pettitt has stated that the *"deliberate placement of flowers has now been convincingly eliminated"*, noting that *"A recent examination of the microfauna from the strata into which the grave was cut suggests that the pollen was deposited by the burrowing rodent Meriones persicus, which is common in the Shanidar microfauna and whose burrowing activity can be observed today"*. A mushroom was found in the personal effects of *Ötzi the Iceman*, whose body was frozen in the Ötztal Alps for more than 5,000 years. The mushroom was probably used to treat whipworm.

ANCIENT TIMES

In ancient Sumeria, hundreds of medicinal plants including myrrh and opium are listed on clay tablets. The ancient Egyptian Ebers Papyrus lists over 800 plant medicines such as aloe, cannabis, castor bean, garlic, juniper, and mandrake.

From ancient times to the present, Ayurvedic medicine as documented in the Atharva Veda, the Rig Veda and the Sushruta Samhita has used hundreds of pharmacologically active herbs and spices such as turmeric, which contains curcumin. The Chinese pharmacopoeia, the *Shennong Ben Cao Jing* records plant medicines such as chaulmoogra for leprosy, ephedra, and hemp. This was expanded in the Tang Dynasty *Yaoxing Lun*.

In the fourth century BC, Aristotle's pupil Theophrastus wrote the first systematic botany text, *Historia plantarum*. In the first century AD, the Greek physician Pedanius Dioscorides documented over 1000 recipes for medicines using over 600 medicinal plants in *De materia medica*; it remained the authoritative reference on herbalism for over 1500 years, into the seventeenth century.

MIDDLE AGES

In the Early Middle Ages, Benedictine monasteries preserved medical knowledge in Europe, translating and copying classical texts and maintaining herb gardens. Hildegard of Bingen wrote *Causae et Curae* ("Causes and Cures") on medicine.

In the Islamic Golden Age, scholars translated many classical Greek texts including Dioscorides into Arabic.

Herbalism flourished in Baghdad and in Al-Andalus. Abulcasis (936–1013) of Cordoba wrote *The Book of Simples*, and Ibn al-Baitar (1197–1248) recorded hundreds of medicinal herbs such as *Aconitum*, nux vomica, and tamarind in his *Corpus of Simples*. Avicenna included many plants in his 1025 *The Canon of Medicine*. Abu-Rayhan Biruni, Ibn Zuhr, Peter of Spain, and John of St Amand wrote further pharmacopoeias.

EARLY MODERN

The early modern period saw the flourishing of illustrated herbals across Europe, starting with the 1526 *Grete Herball*. John Gerard wrote his famous *The Herball or General History of Plants* in 1597, based on Rembert Dodoens, and Nicholas Culpeper published his *The English Physician Enlarged*. Many new plant medicines arrived in Europe as products of Early Modern exploration and the resulting Columbian Exchange. In Mexico, the sixteenth century *Badianus Manuscript* described medicinal plants available in Central America.

PIANTE OFFICINALI

Una **pianta officinale** è un organismo vegetale usato nelle officine farmaceutiche per la produzione di specialità medicinali. Sono considerate piante officinali piante medicinali, aromatiche e da profumo inserite negli elenchi specifici e nelle farmacopee dei singoli paesi. Il numero e il tipo di piante officinali varia da paese a paese a seconda delle tradizioni. Il più comune utilizzo di piante officinali è quello di correttori del gusto: molti farmaci o preparati farmaceutici hanno originariamente un gusto sgradevole, che quindi viene "corretto" con l'aggiunta di sostanze di origine vegetale. Le piante officinali, ad esempio, sono quelle usate per conferire a sciroppi o a caramelle il gusto di fragola, arancia, limone, ecc.

Nel linguaggio comune spesso si sovrappone l'uso dei termini pianta medicinale con pianta officinale, termini che legalmente indicano due diverse entità; il termine officinale è un termine esclusivamente procedurale e indica quelle piante inserite all'interno di elenchi ufficiali come utilizzabili dalle officine farmaceutiche, a prescindere dal fatto che queste piante abbiano o meno proprietà di tipo medicinale. Il termine pianta medicinale indica invece quelle piante che contengono sostanze utilizzabili direttamente a scopo terapeutico o come precursori in emisintesi che portino a sostanze attive. È quindi chiaro che una pianta può essere officinale in un paese e non in un altro, a seconda delle regolamentazioni, ma essa sarà una pianta medicinale a prescindere dalle leggi. Una **pianta medicinale**, secondo l'Organizzazione Mondiale della Sanità (OMS), è un organismo vegetale che contiene in uno dei suoi organi sostanze che possono essere utilizzate a fini terapeutici o che sono i precursori di emisintesi di specie farmaceutiche.

Si va sempre più affermando il concetto di *fitocomplesso*, quale insieme di sostanze di origine vegetale non riproducibili per sintesi chimica. Il fitocomplesso va inteso come l'insieme di una quantità di principi attivi, noti e non, farmacologicamente attivi, e di sostanze che aiutano l'azione dei primi, pur essendo di per sé queste ultime farmacologicamente inattive. L'insieme delle interazioni dei primi (i principi attivi) e dei secondi (i coadiuvanti) determina le azioni note del fitocomplesso. Sempre secondo l'OMS circa il 25% dei moderni farmaci usati in USA sono di origine vegetale; inoltre sono 7.000 circa i composti medici, presenti nella moderna farmacopea, derivati da piante.

PIANTA MEDICINALE E OFFICINALE

Nel linguaggio comune si sovrappone l'uso dei termini pianta medicinale con pianta officinale, che indica piante utilizzate nelle officine farmaceutiche per la produzione di specialità medicinali. Questa definizione è però abbastanza riduttiva, e l'utilizzo in ambienti accademici del termine pianta medicinale non fa più riferimento esclusivamente ad un utilizzo a scopo terapeutico delle sostanze contenute nelle piante, bensì dell'utilizzo della pianta o di estratti da essa derivati a scopo terapeutico.

PIANTE IN PERICOLO

Un'indagine scientifica internazionale promossa dall'OMS all'inizio degli anni novanta, ha rilevato un numero di circa sessantamila specie vegetali, utilizzabili per la cura delle malattie, in forte pericolo di estinzione, di cui trecentosettantaquattro in Italia. Questo fatto richiede una maggiore attenzione alle piante medicinali, non solo quelle utilizzate nelle emisintesi, ma anche quelle che forniscono naturalmente componenti attivi applicabili nell'ambito della fitoterapia.

CENNI STORICI

Con l'introduzione dell'agricoltura si rese necessaria una maggiore attenzione alla vita delle piante e questo fu il punto di partenza della conoscenza, anche medica, delle caratteristiche delle piante stesse. Il più antico documento medico, per ora rintracciato, è il *"papiro di Ebers"*, risalente al 1500 a.C. Gli egizi facevano largo uso di medicamenti di natura vegetale, in particolar modo conoscevano le proprietà della maggiorana, dell'edera, della mirra.

Nell'antica Grecia, le conoscenze sulle piante si mescolarono con le teorie filosofiche sulle stesse. Uno dei più importanti studiosi fu Eracleide, il quale sperimentò nuove ricette, riprese in seguito da Celso. Le radici studiate e messe in vendita vennero definite *"farmacopoli"* e si basavano soprattutto sulle nozioni contenute nei testi medici scritti da Ippocrate (V secolo a.C.) e in quelli botanici scritti da Teofrasto.

Nell'antica Roma, già nel I secolo d.C. vennero impiantati orti chiamati medicinali, in quanto si coltivavano piante sfruttate per le varie terapie mediche.

Nel IX secolo d.C., in Sicilia, grazie ai Saraceni furono introdotte nuove tecniche idrauliche e di irrigazione che consentirono l'introduzione di nuove piante officinali. Gli arabi diedero un grande impulso sia all'alchimia sia alla chimica, che ebbe ripercussioni nello sviluppo farmaceutico di tinture e distillati. Gli arabi furono i primi ad organizzare una farmacopea, quindi un elenco di ricette descriventi le proporzioni e le composizioni chimiche. Ai secoli XI, XII, XIII, risalgono i primi testi farmaceutici, in cui confluirono le influenze greche, romane e arabe, sintetizzate nella definizione delle operazioni fondamentali: lozione, decozione, infusione e triturazione. In questo periodo si diffuse l'uso delle spezie e delle droghe e la Scuola salernitana introdusse assieme alle pratiche chirurgiche anche un antesignano dell'anestesia, la *spongia sonnifera*, imbevuta di oppio, succo di mandragora e di giusquiamo che doveva essere aspirata dal paziente. La Scuola di Salerno si distinse anche per la grande perizia nel selezionare le erbe, sulle quali abbondano indicazioni terapeutiche che si sono dimostrate efficaci ancora ai nostri tempi, valga per tutte l'insegnamento che diceva: «*Purga l'isopo dalle flemme il petto*», che ha un'azione benefica sulle bronchiti e sulle affezioni respiratorie.

La botanica intesa come scienza nacque solo agli inizi del Cinquecento, grazie alle scoperte geografiche e alla introduzione della stampa. Si diffusero, in questo periodo i primi erbari secchi e nel 1533 a Padova fu istituita la prima cattedra di "botanica sperimentale".

Pietro Andrea Mattioli redasse nel 1554 il più significativi testo di medicina e di botanica dell'epoca. Nel Seicento Pierre Magnol inserì nella classificazione l'intuizione delle famiglie, suddividendo il mondo vegetale in settantasei gruppi.

Nel secolo successivo una grande spinta al progresso della botanica fu effettuata dallo svedese Carl von Linné, che identificò le specie viventi dividendole in basi alle classi, agli ordini e ai generi.

Da allora l'evoluzione è stata continua.

I
THE PLATES
LE TAVOLE

PLANTS, MEDICINAL

HAND-COLORED ENGRAVING BY

JAMES SOWERBY

PUBLISHED BY

DR. WOODVILLE.

PLATES LIST OF ILLUSTRATIONS

1. Pinus sylvestris - Scotch fir
2. Pinus abies - Norway spruce fir tree
3. Pinus picea - Silver fir tree
4. Pinus larix - Common white larch tree
5. Juniperus sabina - Common savin
6. Juniperus communis - Commun Juniper
7. Juniperus lycia - Lycian Juniper or Cedar
8. Salix fragilis - Crack willow
9. Juglans regia - Common walnut tree
10. Quercus robur - Common oak
11. Pistacia lentiscus - Mastich tree
12. Pistacia terebinthus - Chian or Cyprus turpentine tree
13. Arctium lappa - Burdock
14. Centaura Benedicta - Blessed or Holy Thistle
15. Achillea Millefolium - Common Yarrow or Milfoil
16. Leontodon Taraxacum - Common Dandelion
17. Arnica Montana - Mountain Arnica
18. Tussilafo Farfara - Coltsfoot
19. Anthemis Nobilis - Common Camomile
20. Anthemis Pyrethrum - Spanish Camomile or Pellitory of Spain
21. Artemisia Abrotanum - Common Southernwood
22. Artemisia Absinthium - Common Wormwood
23. Artemisia Vulgaris - Mugwort
24. Artemisia Maritima - Sea Wormwood
25. Artemisia Santonica - Tartarian Southernwood
26. Inula Helenium - Common Inula or Elecampane
27. Tanacetum Vulgare - Common Tansy
28. Cynara Scolymus - Common Artichoke
29. Cichorium Intybus - Wild or Blue Succory
30. Matricaria Parthenium - Common Feverfew
31. Lactuca Virosa - Strong-scented wild Lettuce
32. Valeriana Officinalis - Officinal Valerian
33. Plantago Major- Common great Plantane or Way-bread
34. Viscum Album - Missletoe
35. Angelica Archangelica - Garden Angelica
36. Angelica Sylvestris - Wild Angelica
37. Phellandrium Aquaticum - Fine leaved water-hemlock
38. Oenanthe Crocata - Hemlock water-dropwort
39. Cicuta Virosa - Water Hemlock
40. Bubon Galbanum - Lovage-leaved Bubon
41. Carum Carui - Common Caraway
42. Conium Maculatum - Common Hemlock
43. Ferula Assafoetida - Assafoetida Gigantic Fennel
44. Imperatoria Ostruthium - Common Masterwort
45. Apium Petroselinum - Common Parsley
46. Eryngium Maritimum - Sea eryngo or Holly
47. Pastinaca Opopanax - Opopanax or Rough Parsnep

48. Anethum Graveolens - Common Dill
49. Anethum Foeniculum - Common Fennel
50. Daucus Carota - Wild Carrot or Bird's Nest
51. Pimpinella Saxifraga - Small Burnet-Saxifrage
52. Pimpinella Anisum - Anise
53. Coriandrum Sativum - Common Coriander
54. Sium Nodiflorum - Creeping water Parsnep
55. Ligusticum Levisticum - Common Lovage
56. Cuminum Cyminum - Cummin
57. Vitis Vinifera - Common Vine
58. Panax quinquefolium - Ginseng
59. Aristolochia Serpentaria - Snake-rooted birthwort
60. Aristolochia Longa - Long-rooted birthwort
61. Aristolochia Clematitis - Climbing birthwort
62. Smilax Sarsaparilla
63. Smilax China - Chinese Smilax
64. Ruscus aculeatus - Butcher's broom or Knee Holly
65. Cissampelos Pareira - Pareira Brava Cissampelos
66. Asarum Europaeum - Common Asarabacca
67. Rubia Tinctorum - Dier's Madder
68. Galium Aparine - Cleavers or Goose grass
69. Spigelia Marilandica - Perennial Worm Grass or Indian pink
70. Coffea Arabica - Cofee tree
71. Cucumis Colocynthis - Bitter Cucumber or Coloquintida
72. Momordica Elaterium - Wild or squirting Cucumber
73. Bryonia Alba - White Bryony
74. Datura Stramonium - Common thorn-apple
75. Verbascum Thapsus - Great broad-leaved Mullein
76. Hyoscyamus Niger - Black Henbane
77. Nicotiana Tabacum - Virginian Tobacco
78. Digitalis Purpurea - Common Fox-glove
79. Strychnos Nux Vomica - Vomic Nut or poison Nut
80. Capsicum Annuum - Annual Capsicum or Guinea Pepper
81. Physalis Alkekengl - Common winter Cherry
82. Atropa Belladonna - Deadly Nightshade
83. Atropa Mandragora - Mandrake
84. Solanum Nigrum - Garden Nightshade
85. Solanum Duclamara - Woody Nightshade
86. Convolvulus Scammonia - Scammony bind-weed
87. Convolvulus Jalapa - Jalap bind-weed
88. Lobelia Siphilitica - Blue Lobelia or Cardinal Flower
89. Viola Odorata - Sweet Violet
90. Viola Tricolor - Pansie or three-coloured Violet
91. Cinchona Officinalis - Official Cinchona or Peruvian bark Tree
92. Cinchona-cortex Peruvianus Ruber - Red Peruvian bark Tree
93. Asclepias Vincetoxicum - Official Swallow-wort
94. Gentiana Purpurea - Purple Gentian
95. Gentiana Lutea - Yellow Gentian
96. Chironia Centaurium - Centaury
97. Menyanthes Trifoliata - Water Trifoil or Buckbean
98. Olea Europaea - European Olive

Pinus sylvestris

Published by Phillips & Fardon, Jan.1st 1805.

1 - *Pinus sylvestris* - *Scotch fir*
Order Coniferae

Pinus Abies

2 - *Pinus abies - Norway spruce fir tree*
Order Coniferae

Pinus Picea

Published by Phillips & Farden, Jan.1.st 1805.

3 - Pinus picea - Silver fir tree
Order Coniferae

Pinus Larix

Published by Phillips & Purdon, Jan.t 1.st 1812.

4 - *Pinus larix- Common white larch tree*
Order Coniferae

Juniperus Sabina

Published by Phillips & Burden Jan 1 ...

5 - Juniperus sabina - Common savin
Order Coniferae

6

Juniperus communis

Published by Phillips & Farden, Feb 12th 1805.

6 - *Juniperus communis* - *Commun juniper*
Order Coniferae.

Juniperus Lycia

Published by Phillips & Fardon, Feb.1st 1805.

7-Juniperus lycia - Lycian juniper or Cedar
Order Coniferae - Olibanum

8- Salix fragilis - Crack willow
Order Amentaceae

9-Juglans regia - Common walnut tree
Order Amentaceae

Quercus Robur

Published by Phillips & Borden. Feb 1st 1803.

10- Quercus robur - Common oak
Order Amentaceae

Pistacia Lentiscus

Published by Phillips & Fardon, March 1803.

11- *Pistacia lentiscus - Mastich tree*

Order Amentaceae

Pistacia Terebinthus

Published by Wilkes & Norden, March 1st 1800.

12- *Pistacia terebinthus - Chian or Cyprus turpentine tree*
Order Amentaceae - Ex qua fiuit Terebinthina chia.

13- Arctium lappa - Burdock
Order Compositae

Centaurea Benedicta

Published by Phillips & Farrior, May ...

14- *Centaura Benedicta - Blessed or Holy Thistle*
Order Compositae

Achillea Millefolium.

15- *Achillea Millefolium - Common Yarrow or Milfoil*
Order Compositae

Leontodon   *Taraxacum*

Published by Phillips ... Garden. April 1. 1806

16- *Leontodon Taraxacum - Common Dandelion*
Order Compositae

Arnica *montana*

Published by Phillips & Forden, April 1st 1806.

17- *Arnica Montana - Mountain Arnica*

Order Compositae

18- Tussilafo Farfara - Coltsfoot
Order Compositae

Anthemis nobilis

Published by Phillips & Purdon, April 1st 1806.

19- *Anthemis Nobilis - Common Camomile*

Order Compositae

Anthemis Pyrethrum

Published by Phillips & Fardon, April 1st 1806.

20- *Anthemis Pyrethrum - Spanish Camomile or Pellitory of Spain*
Order Compositae

Artemisia Abrotanum

Published by Phillips & Parker, May 1st 1818

Artemisia Absinthium

Published by Phillips & Parker, May 1st 1818

Artemisia vulgaris

Published by Phillips & Parker, May 1st 1818

Artemisia maritima

Published by Phillips & Parker, May 1st 1818

21-22-23-24- *Anthemisia Various*

Order Compositae

Artemisia Santonica

Published by Phillips & Jardon, May 1 1806.

25- *Artemisia Santonica - Tartarian Southernwood*
Order Compositae

Inula Helenium

Published by Phillips & Fardon, June 1st 1808.

26 - *Inula Helenium - Common Inula or Elecampane*
Order Compositae

Tanacetum vulgare.

Published by Phillips & Yardon, June 1st 1806

27- *Tanacetum Vulgare - Common Tansy*
Order Compositae

Cynara Scolymus

Published by Phillips & Forden, June 1.st 1808.

28- Cynara Scolymus - Common Artichoke
Order Compositae

Cichorium Intybus
Published by Phillips & Parker, June 1st 1806.

29- *Cichorium Intybus - Wild or Blue Succory*
Order Compositae

30- *Matricaria Parthenium - Common Feverfew*
Order Compositae

31.

Lactuca virosa

Published by Phillips, & Farden, July 1st 1808.

31- *Lactuca Virosa - Strong-scented wild Lettuce*
Order *Compositae*

Valeriana officinalis

Published by Phillips, & Fardon, July 1st 1808.

32- *Valeriana Officinalis - Officinal Valerian*
Order Aggregatae

Plantago major

Published by Phillips & Pardon. July 1.st 1808.

33- *Plantago Major- Common great Plantane or Way-bread*
Order Conglomeratae

Viscum album.

Published by Phillips, & Pardon, July 1st 1818.

34- *Viscum Album - Missletoe*

Order Conglomeratae

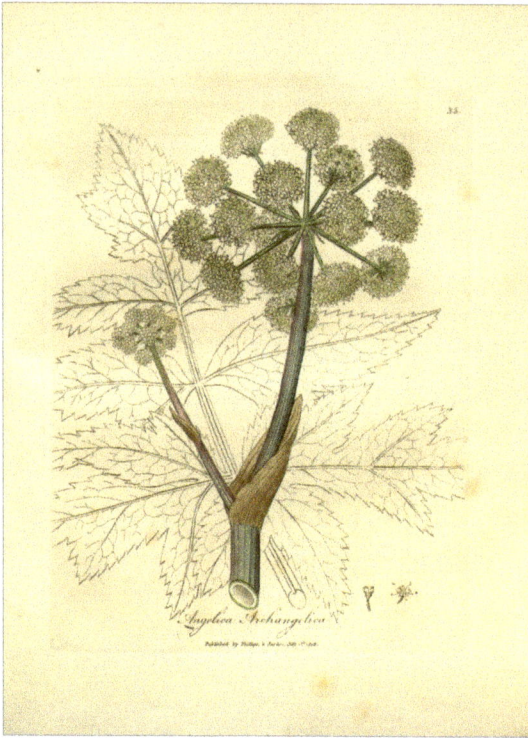

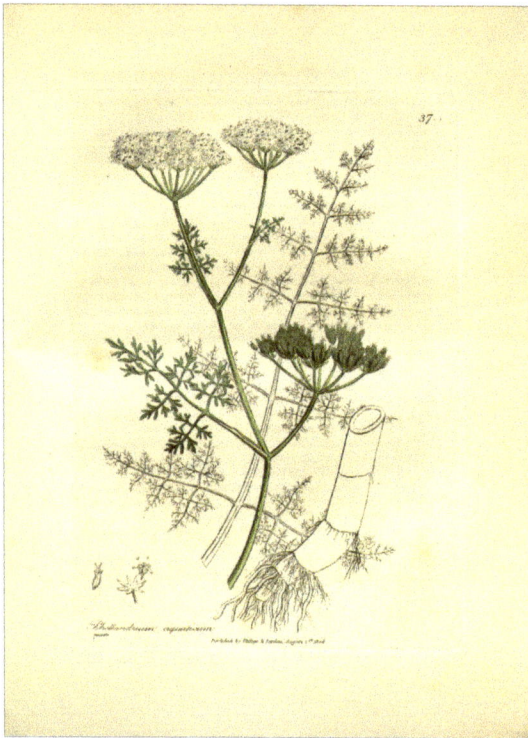

35-36-37-38- Umbrellatae various
Order Umbrellatae

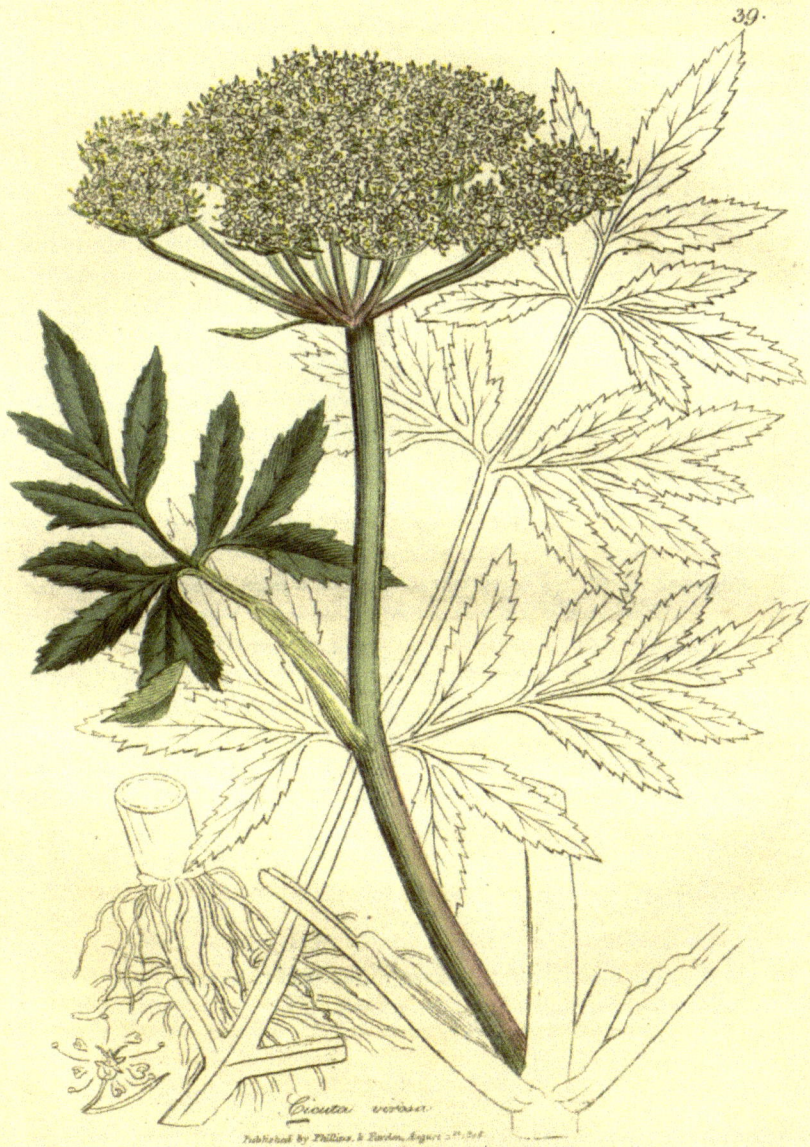

Cicuta virosa.

Published by Phillips, & Purton, August 1st 1818.

39- Cicuta Virosa - Water Hemlock
Order Umbrellatae

Bubon Galbanum.

Published by Phillips, & Fardon, August 1st 1806.

40- *Bubon Galbanum - Lovage-leaved Bubon*
Order Umbrellatae

41- *Carum Carui - Common Caraway*
Order Umbrellatae

Conium maculatum

Published by Phillips & Fardon, Sep.r 1.st 1806.

42- *Conium Maculatum - Common Hemlock*
Order Umbrellatae

Ferula Assafœtida

Published by Phillips & Farden, Sep.t 1st 1808.

43- *Ferula Assafoetida- Assafoetida Gigantic Fennel*
Order Umbrellatae

Imperatoria Ostruthium

Published by Phillips & Fardon June 1 1806

44- *Imperatoria Ostruthium* - *Common Masterwort*

Order Umbrellatae

Apium Petroselinum
45.

Pastinaca Opopanax
47.

Anethum graveolens
48.

Anethum Foeniculum
49.

45-47-48-49- *Apium and Anrthum various*
Order Umbrellatae

Eryngium maritimum

Published by Phillips, & Fardon, Oct.ʳ 1ˢᵗ 1806.

46- *Eryngium Maritimum - Sea eryngo or Holly*
Order Umbrellatae

50

Daucus Carota

Published by Phillips & Fardon, Oct 1st 1806.

50- *Daucus Carota* - *Wild Carrot or Bird's Nest*
Order Umbrellatae

Pimpinella Saxifraga

Published by Phillips & Fardon, Nov. 1st 1806.

51- *Pimpinella Saxifraga - Small Burnet-Saxifrage*
Order Umbrellatae

Sium nodiflorum

Published by Phillips, & Jarden, Nov'r. 1st 1808.

54- *Sium Nodiflorum - Creeping water Parsnep*
Order *Umbrellatae*

Pimpinella Anisum

Coriandrum sativum

Ligusticum Levisticum

Cuminum Cyminum

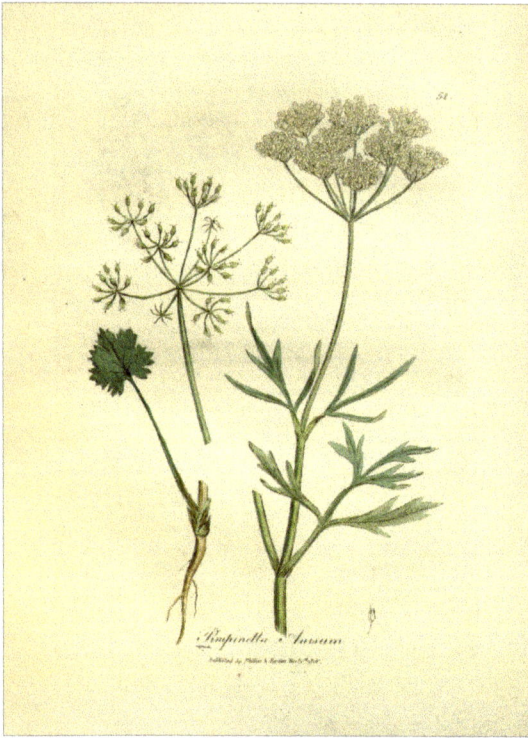

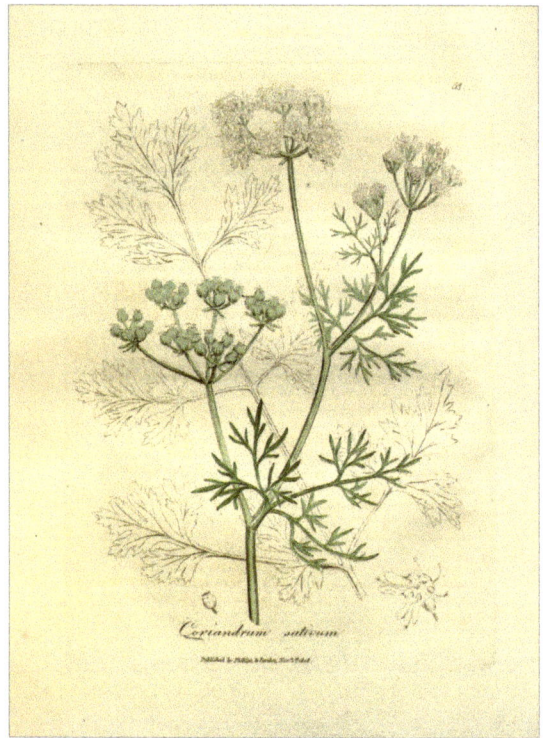

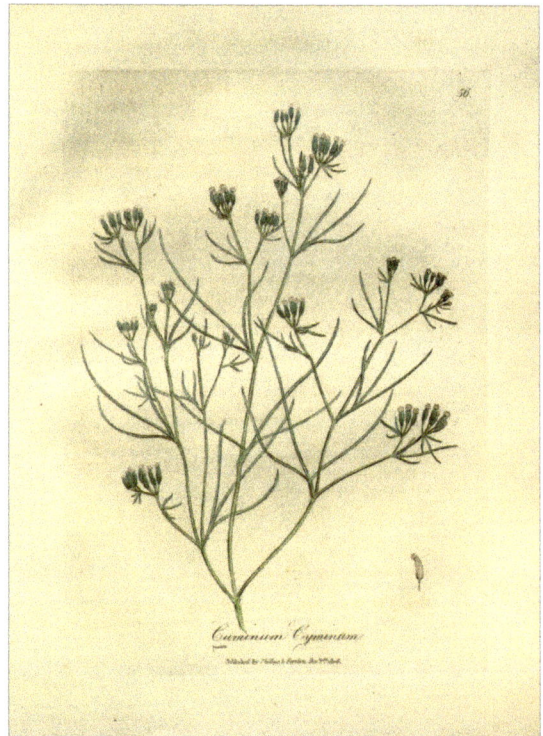

52-53-55-56- *Umbrellatae various*
Order Umbrellatae

Vitis vinifera

Published by Ph. Rips & Harden, Dec.r 1.st 1806.

57- Vitis Vinifera - Common Vine
Order Hederaceae

Panax quinquefolium

Published by Phillips & Farden. Dec 1st 1808.

58- Panax quinquefolium - Ginseng
Order Hederaceae

Aristolochia *Serpentaria*

Published by Phillips, & Fardon, Dec.r 1.st 1808.

59- *Aristolochia Serpentaria* - *Snake-rooted birthwort*

Order Sarmentaceae

Aristolochia longa

Published by Phillips, & Fardon, Dec.3st 1806.

60- *Aristolochia Longa* - *Long-rooted birthwort*

Order Sarmentaceae

Aristolochia Clematitis.

Published by Phillips & Parsons, Jan'y 1st 1807.

61- *Aristolochia Clematitis - Climbing birthwort*

Order Sarmentaceae

Smilax Sarsaparilla.

Published by Phillips & Parker, Jan.ry 1st 1807.

62- Smilax Sarsaparilla

Order Sarmentaceae

Smilax China.

Published by Phillips & Fardon, June 1 1813.

63- *Smilax China - Chinese Smilax*
Order Sarmentaceae

64- Ruscus aculeatus - Butcher's broom or Knee Holly

Order Sarmentaceae

Cissampelos Pareira
Published by Phillips & Forder, Jan.y 1.st 1807.

65- *Cissampelos Pareira - Pareira Brava Cissampelos*
Order Sarmentaceae

Asarum europæum

Published by Phillips & Yarden, Feby. 1st, 1807.

66- *Asarum Europaeum - Common Asarabacca*
Order Sarmentaceae

Rubia tinctorum

Published by Phill....

67- *Rubia Tinctorum - Dier's Madder*

Order Stellatae

Galium Aparine

Published by Phillips & Fardon, Feby 1st 1807.

68 - *Galium Aparine - Cleavers or Goose grass*
Order Stellatae

Spigelia marilandica

Published by Phillips & Pardon, Feb'y 1st 1807.

69- *Spigelia Marilandica - Perennial Worm Grass or Indian pink*
Order Stellatae

70- *Coffea Arabica* - *Cofee tree*
Order Cymosae

Cucumis Colocynthis

Publish'd by Phillips & Fardon, Sharbes Valley

71- Cucumis Colocynthis - Bitter Cucumber or Coloquintida
Order Cucurbitaceae

Momordica Elaterium

72- *Momordica Elaterium - Wild or squirting Cucumber*
Order Cucurbitaceae

Bryonia alba

Published by Phillip & Co. 1st March 1807

73- *Bryonia Alba - White Bryony*
Order Cucurbitaceae

Datura Stramonium

Published by Phillips & Fardon, March 1st 1807.

74- Datura Stramonium - Common thorn-apple

Order Solenaceae, seu Luridae

75- *Verbascum Thapsus - Great broad-leaved Mullein*
Order Solenaceae, seu Luridae

Hyoscyamus niger

Published by Phillips & Fardon, April 1st 1809.

76- Hyoscyamus Niger - Black Henbane
Order Solenaceae, seu Luridae

Nicotiana Tabacum

Published by Phillips & Farden, April 1st 1807.

77- *Nicotiana Tabacum* - *Virginian Tobacco*
Order Solenaceae, seu Luridae

78-79-80-81- Solanaceae various

Order Solenaceae, seu Luridae

Atropa Belladonna.

Published by Phillips & Fardon, May 1st 1807

82- *Atropa Belladonna - Deadly Nightshade*
Order *Solenaceae, seu Luridae*

83- Atropa Mandragora - Mandrake
Order Solenaceae, seu Luridae

84.

Solanum nigrum

Published by Phillips & Borden, May 1st 1809.

84- *Solanum Nigrum* - *Garden Nightshade*
Order Solenaceae, seu Luridae

Solanum Dulcamara

Published by Phillips & Fardon, May 1st 1807.

85- Solanum Duclamara - Woody Nightshade

Order Solenaceae, seu Luridae

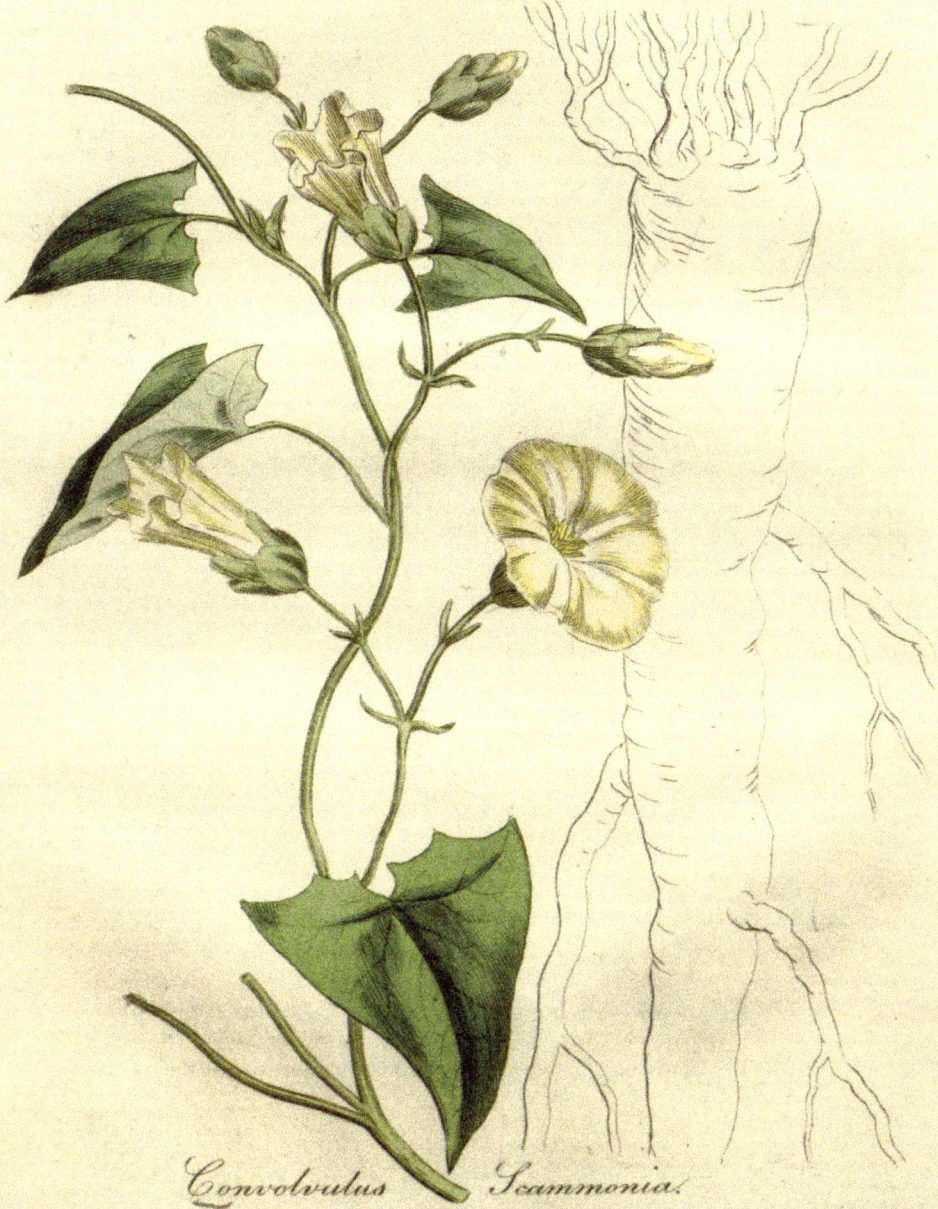

Convolvulus Scammonia.

Published by Phillips & Forden, June 1st 1807.

86- *Convolvulus Scammonia - Scammony bind-weed*
Order Campanaceae

Convolvulus Jalapa.

Published by Phillips & Fardon June 1st 1807.

87- Convolvulus Jalapa -Jalap bind-weed
Order Campanaceae

Lobelia siphilitica

88- Lobelia Siphilitica - Blue Lobelia or Cardinal Flower
Order Campanaceae

Viola odorata

Published by Phillips & Darton, Jan 1st 1807.

89- *Viola Odorata* - *Sweet Violet*

Order Campanaceae

Viola tricolor

Published by Philips & Fardon, June 1st 1807.

90- *Viola Tricolor - Pansie or three-coloured Violet*
Order Campanaceae

91- Cinchona Officinalis - Officinal Cinchona or Peruvian bark Tree
Order Contortae

Cinchona Cortex peruvianus ruber

Published by Phillips & Fardon, July 1ˢᵗ 1807.

92- *Cinchona-cortex Peruvianus Ruber - Red Peruvian bark Tree*

Order Contortae

Asclepias Vincetoxicum.

93- *Asclepias Vincetoxicum - Officinal Swallow-wort*
Order Contortae

Gentiana purpurea

Published by Phillips & Furden, July 1st 1809.

94- *Gentiana Purpurea - Purple Gentian*
Order Rotaceae

Gentiana lutea

Published by Philips & Fardon, July 1st 1807

95- *Gentiana Lutea - Yellow Gentian*
Order *Rotaceae*

Chironia Centaurium

Published by Philips & Fardon, August 1st 1807.

96- Chironia Centaurium - Centaury
Order Rotaceae

Menyanthes trifoliata.

97- *Menyanthes Trifoliata - Water Trifoil or Buckbean*

Order Rotaceae

Olea europæa

Published by Phillips & Fardon, August 1st 1809.

98- Olea Europaea - European Olive
Order Sepeariae

DARWIN'S VIEW SERIES

Actually, the world from Darwin's point of view. The new series of specifically dedicated to the animal, vegetable and mineral world. A great review of nature through his most beautiful and fascinating images, taken from ancient tomes and essays about nature, made by the greatest individuals, artists and scientists together. Not only that, "Darwin's view" will involve yourself through the description of the stories, with facts and images of the exotic and romantic travels, made by the great explorers and brilliant scientists of the past, starting with the epic one on the HMS Beagle of our beloved and legendary Charles Robert Darwin!

CONTENTS

PREFACE   PAG. 5

BIOGRAPHIES   PAG. 7

MEDICAL PLANTS   PAG. 8

PIANTE OFFICINALI (ITALIAN TEXT)   PAG. 10

PLATES/TAVOLE   PAG. 13

MUS
EUM
BIBLIOTHECA

www.ingramcontent.com/pod-product-compliance
Lightning Source LLC
Chambersburg PA
CBHW051915210326
41597CB00033B/6156